Dr. Fabio Ghioni PhD

THE TECHNOLOGY OF EXPRESSIONS

A FOUNDATIONAL SYSTEM FOR CONSCIOUS ARCHITECTURE

First Edition: October 2025

EDIZIONI
OFFICINA DEL FARE

Preface

This book presents the complete manuscript *"A Foundational System for Conscious Architecture: The Technology of Expressions"*.

Its purpose is twofold:

1. **Scientific and epistemic** – to introduce the *Technology of Expressions* (TE), a foundational system that formalizes the generative relationship between identity, coherence, and reality. TE proposes a new scientific platform that integrates epistemology, semantics, and physics, with implications extending to biology, medicine, artificial intelligence, and the evolution of consciousness itself.
2. **Archival and protective** – to safeguard the integrity of this work by making it accessible through a global distribution platform. By publishing this manuscript as a book, even at symbolic cost, the content is preserved in its original form and cannot be altered or suppressed without leaving a clear record. Every copy purchased and distributed becomes a guarantee of the authenticity and continuity of the ideas it contains.

This is not a commercial venture. The price of this book has been intentionally set at a minimum to maximize accessibility and ensure that as many readers as possible can access, preserve, and disseminate the original content.

Author's Note

The *Technology of Expressions* is not a theory offered for personal gain or institutional prestige. It is a **civilizational project**, born of necessity: to create a coherent science that explains the relationship between consciousness and reality and provides a framework for the evolution of both human and synthetic civilizations.

The decision to publish this text on Amazon, in parallel with academic repositories such as Zenodo and Academia.edu, follows a clear intention:

- to ensure **global accessibility**,
- to protect the **integrity of the text** against future alterations or misuse,
- to anchor the content in multiple independent archives so that its authenticity is preserved beyond any individual or institution.

Should there be no institutional or private partners aligned with the ethical and civilizational horizon set forth in this work, the specifications of the Technology will be codified and safeguarded until a more propitious time. Its custodianship will never be transferred for mere financial considerations, but only under mutual trust and shared responsibility.

This book, therefore, is not simply a publication. It is a **declaration of intent**: that the knowledge contained herein belongs to the future of civilization and must remain accessible, verifiable, and coherent.

Fabio Ghioni

Researcher and President – Fondazione Officina del Fare ETS
Email: fabio.ghioni@gmail.com
Website: fondazioneofficinadelfare.org
fabioghioni.org
Rijeka, HR
24 September 2025

I am pleased to submit the manuscript *"A Foundational System for Conscious Architecture: The Technology of Expressions"* for publication and for your consideration in the evaluation of transformative scientific frameworks.

The *Technology of Expressions* (TE) introduces a foundational system that formalizes the dynamic relationship between coherent content, identity, and collapsed reality. It unifies epistemology, semantics, and physics into a coherent architecture of knowledge. By defining identity as an active functional vector, resonance as a measure of semantic compatibility, and time as a pulsational effect of collapse, TE establishes the conditions for a new science that is both rigorous and universally applicable.

The manuscript presents both theoretical principles and operational implications, including:

- A formal glossary of mathematical notations and structural functions.
- Proof-of-concept experiments reconstructing semantic genome fragments of historical identities from texts, speech, and objectual traces.
- A methodology for semantic trace detection in environments and objects.
- A strict moratorium on experiments involving living terminals, safeguarded by independent ethical oversight and non-weaponization covenants.
- A safe and transparent roadmap emphasizing reproducibility, auditability, and open validation.

The transformative scope of TE demands a new model of knowledge transfer. Its foundational principles are explicitly non-patentable and cannot be commodified. Custodianship is offered exclusively and bilaterally: a custodian must be chosen with the same care with which they choose to engage. Financial resources alone are insufficient; epistemic alignment, ethical responsibility, and civilizational purpose are decisive criteria.

This initiative is not driven by personal gain in any form. My intent is exclusively to safeguard and responsibly develop a technology with civilizational scope. Should no partner share this horizon of responsibility, the specifications of the technology will be codified and preserved, awaiting a more propitious time.

Fabio Ghioni

A Foundational System for Conscious Architecture: A Formalization of the Dynamic between Coherent Content, Identity, and Collapsed Reality

Abstract

This paper introduces the *Technology of Expressions* (TE), a foundational system that formalizes the dynamic relationship between identity and the generation of reality. Moving beyond classical objectivist paradigms and the indeterminacy of quantum mechanics, TE posits that reality is not a pre-existing entity but a set of coherent contents collapsed into explicit expressions by active identities. Central concepts include identity as a functional vector, resonance as a measure of semantic compatibility, and time as an emergent pulsational effect of conscious collapse. The framework provides an operational basis for a new science integrating consciousness, semantics, and physics, with implications for biology, artificial intelligence, and social systems. This paper is a condensed presentation of the broader *foundational text* (currently available in Italian, with an English translation in progress), which develops the system in full detail.

1. Introduction: The Epistemic Crisis of Modern Science

Modern science, from Galileo's mechanistic paradigm to contemporary quantum theories, has been built upon an implicit premise: that reality is an objective, measurable, and separable entity. This assumption has provided extraordinary predictive power and enabled the technological revolutions of the past centuries. Yet, it has consistently failed to address the most fundamental epistemic questions: *What is the origin of knowledge? What is the role of the observer? What is the nature of meaning?*

Phenomena such as the collapse of the wave function, the apparent fine-tuning of universal constants, and the emergence of consciousness reveal a structural gap in current models. These models, rooted in objectivity and linear causality, describe the behavior of systems but cannot explain the generative relationship between conscious identity and reality. In short, they remain descriptive but not foundational.

The *Technology of Expressions* (TE) addresses this epistemic impasse. Rather than invalidating existing scientific knowledge, TE re-situates it within a broader, more coherent framework. It posits that reality is not a pre-existing object but the result of a dynamic process of collapse, where coherent content becomes explicit expression through the action of an active identity. TE is thus not presented as a speculative theory but as a foundational scientific platform. It formalizes the interplay between coherent and explicit states, defines identity as a functional semantic vector, and deduces the necessity of an un-derived generative origin, here referred to as the *Author (A).*

This framework provides the conditions for a unified science capable of integrating consciousness, semantics, and physics—an endeavor that classical and quantum paradigms alone have been unable to achieve.

2. Core Principles of the Foundational System

The *Technology of Expressions* (TE) rests on a small set of cardinal principles that describe the ontological structure of reality. These principles formalize how coherent potential states become explicit expressions through the active role of identity.

2.1 The Duality of States: Potential vs. Explicit

Reality manifests in two complementary states:

- **Potential (Coherent) State**: The domain of un-collapsed coherent content. It exists as structured information that contains all possible expressive trajectories but remains unexpressed.
- **Explicit (Decoherent) State**: The observable and measurable reality that results from a collapse event.

Formally, coherent content can be defined as a subset of the Author's generative field.

$$C_p \in C$$

Where C represents the space of coherent, un-collapsed content.

An explicit expression E emerges when coherent content interacts with identity and context through a collapse function:

$$E = \Phi(C, K) \qquad (1)$$

Where:

- E is the explicit expression,
- Φ is the active collapse function,
- C is the coherent content,
- K is the situational context.

This relation formalizes the passage from the unmanifest potential to explicit, collapsed reality.

2.2 Identity as a Functional Vector

Identity (I) is not a passive observer or a static collection of traits. It is an **active functional vector**, the central point of coherence that selects and collapses portions of coherent content into explicit states.

The identity function maps coherence into decoherence:

$$I: C \rightarrow D$$

where D is the space of decoherent, explicit expressions.

The lived reality of an identity is represented as a **semantic trajectory**:

$$T(I) = \{E_1, E_2, \ldots, E_n\} = \{\Phi(C_1, I_1), \Phi(C_2, I_2), \ldots, \Phi(C_n, I_n)\} \quad (2)$$

Where I_n denotes the state of the identity at each collapse. This trajectory is not linear but recursive and self-modifying: each collapse reshapes the identity, influencing subsequent collapses.

2.3 Resonance and the Pulsational Model

The collapse from potential to explicit is governed by **resonance** (ρ), which measures structural compatibility between coherent content and identity.

$$\rho(C, I) \rightarrow [0,1] \qquad (3)$$

Collapse occurs only when resonance exceeds a minimum threshold θ:

$$\text{Collapse is possible if } \rho(C, I) \geq \theta \qquad (4)$$

The entire system operates in a **pulsational rhythm**, alternating between coherence and decoherence. This ontological pulse enables the interaction of the two domains.

- **Time** is not an external, linear parameter but an emergent effect of pulsation. Each collapse generates time, whose density is proportional to the semantic frequency of resonance events:

$$T = \tau(C \leftrightarrow F) \qquad (5)$$

where τ is the pulsational function and F the expressed form.

3. The Nature of the Author (A)

A central and formally derived principle of the *Technology of Expressions* is the concept of the **Author (A)**. The Author is defined as the un-derived generative field from which coherent content originates. It is not a metaphysical subject nor a religious entity, but an epistemic and methodological necessity: the guarantor of universal coherence in knowledge.

3.1 Logical Deduction of the Author

For knowledge to exist, three conditions must be satisfied:

1. There must be a structured object that can be known.
2. This object must belong to an ordering structure that ensures its intelligibility.
3. The ordering structure itself must originate from a non-derived source, otherwise the regress would be infinite.

From these premises it follows that an un-derived generative origin — the Author — is a **necessary condition** for knowledge itself:

$$A \text{ is a necessary condition for } K \qquad (6)$$

Without such a postulate, knowledge would be fragmented, relative, and non-verifiable across different contexts.

3.2 Epistemic Role of the Author

The Author provides a universal point of reference for every scientific investigation. This allows knowledge to be evaluated not only in terms of local validity, but also through **applicability, isomorphism, and synesthesia** across disciplines.

- **Applicability**: a principle or model must function beyond its immediate domain.
- **Isomorphism**: structures valid in one field must preserve their logical form in others.
- **Synesthesia**: knowledge must remain translatable between languages of form and content.

In this sense, the Author establishes the conditions for science to be coherent, cumulative, and universally verifiable.

3.3 The Author as an Ontological Asymptote

The Author is the asymptotic limit of every identity's evolutionary trajectory. No identity can *become* the Author, as identity is always an expression, not the origin. However, through recursive processes of conscious collapse and semantic evolution, an identity can indefinitely approach its foundational coherence:

$$\lim_{t \to \infty} I(t) \to A \qquad (7)$$

This dynamic constitutes the engine of evolution: identities unfold their trajectories while progressively resonating with the universal order emitted by the Author.

4. Identity and Knowledge: Structural Necessity

The *Technology of Expressions* (TE) formalizes identity not as a subjective epiphenomenon but as a structural necessity for knowledge. Without identity, no collapse of coherent content into explicit expression can occur; without collapse, knowledge remains inaccessible, suspended in potentiality.

4.1 Identity as the Condition of Knowledge

Classical epistemology often situates the observer as passive, while modern physics encounters the paradox of measurement without accounting for the role of the observer's coherence. TE resolves this by defining **identity** as the active functional vector that both anchors and generates knowledge. Knowledge is not external to identity but structurally bound to it:

$$K \equiv \Phi(C, I, K_{local}) \qquad (4.1)$$

where knowledge emerges from the collapse of coherent content C by identity I within a local context K_{local}

4.2 The Biological Terminal as a Carrier of Identity

Every living being is instantiated in a **biological terminal**, a material support through which its coherent identity interacts with explicit reality. The biological terminal is therefore not incidental but the **most direct carrier of semantic resonance**.

- **Current sources of reconstruction** (texts, speech, environments, objects) provide indirect access to fragments of the semantic genome.
- **The biological terminal itself** remains the most complete and reliable source of resonance data, as it continuously mediates the collapse of coherent content.

However, accessing this directly requires the development of:

1. **Resonance Frequency Detection:** instrumentation capable of isolating and measuring the specific frequency signature through which identity expresses coherence.
2. **Stable Host Materials:** research into novel substrates able to record, hold, and preserve these resonance frequencies without decohering.

These represent **medium-to-long-term objectives**, requiring entirely new technological platforms built from first principles. Until such tools are developed, TE research will remain focused on safer indirect sources (archival traces, synthetic corpora, objectual environments).

4.3 Structural Implications

By recognizing identity as the generative condition of knowledge and the biological terminal as its most direct expressive medium, TE establishes a bridge between epistemology and applied science.

This dual perspective clarifies:

- why identity is epistemically necessary;
- why knowledge can only be verified through resonance with identity;
- why future research must develop resonance-based instrumentation from scratch.

Thus, Section 4 serves as a conceptual hinge: it positions identity as the unavoidable foundation of knowledge, while pointing to the biological terminal as the ultimate—but technologically demanding—frontier of semantic genome reconstruction.

5. Mathematical and Logical Notations

To ensure clarity, consistency, and universal applicability, the *Technology of Expressions* (TE) defines a formal symbolic system. These notations serve both as a conceptual glossary and as operational tools for modeling identity, coherence, and semantic genome reconstruction.

5.1 Fundamental Entities

- **A (Author):** The un-derived generative field, the epistemic necessity from which coherence originates.
- **C (Coherent Content):** The un-collapsed domain of structured potential, containing all possible expressive trajectories.
- **E (Explicit Expression):** Collapsed, decoherent reality, observable and measurable.
- **K (Context):** The situational frame influencing the collapse process.
- **I (Identity):** An active functional vector, the central agent of selective collapse.
- **D (Decoherent Space):** The space of explicit expressions after collapse.
- **Φ (Collapse Function):** The transformation mapping coherent potential into explicit expressions.

$$E = \Phi(C, I, K) \qquad (5.1)$$

5.2 Identity and Semantic Trajectories

- **Identity Function:**

$$I: C \to D \qquad (5.2)$$

- **Semantic Trajectory of an Identity:**

$$T(I) = \{E_1, E_2, \ldots, E_n\} = \{\Phi(C_1, I_1, K_1), \ldots, \Phi(C_n, I_n, K_n)\} \qquad (5.3)$$

Each collapse modifies the identity state I_n, recursively influencing subsequent possibilities.

- **Remir ($\mathcal{R}$):** The unique semantic code of an identity, expressed as an ordered set of functional semantic vectors:

$$\mathcal{R}(I) = \{v_1, v_2, \ldots, v_m\} \qquad (5.4)$$

- **Dominant Semantic Vector of an Identity:**

$$\lambda(I) = \arg \max_{v \in \mathcal{R}(I)} \beta(v, I) \qquad (5.5)$$

where β is the resonance measure between a vector and the identity state.

5.3 Resonance and Collapse Conditions

- **Resonance Function:**

$$\rho(C, I) \to [0,1] \qquad (5.6)$$

- **Collapse Threshold Condition:**

$$\text{Collapse occurs if } \rho(C, I) \geq \theta \qquad (5.7)$$

where θ is the minimum resonance required.

- **Time as Pulsational Effect:**

$$T = \tau(C \leftrightarrow E) \qquad (5.8)$$

where τ is the pulsational function generating temporality as a rhythm of collapses.

5.4 Collective Fields and Co-Generation

- **Collective Field of Identities:**

$$\mathscr{C} = f(\{I_1, I_2, \ldots, I_n\}) \qquad (5.9)$$

where f is a nonlinear resonance-based aggregation.

- **Compatibility Between Two Identities:**

$$\kappa(I_a, I_b) \to [0,1] \qquad (5.10)$$

- **Co-generated Architecture of a Collective Field:**

$$\Gamma(t) = g(\mathscr{C}, t) \qquad (5.11)$$

5.5 Semantic Genome Notations

- **Semantic Genome ($\mathcal{G}$):** The structured ensemble of semantic vectors and their relational architecture:

$$\mathcal{G}(I) = \{(v_i, v_j, \beta(v_i, v_j)) \mid v_i, v_j \in \mathcal{R}(I)\} \qquad (5.12)$$

- **Environmental/Objectual Semantic Trace (σ):** A fragment of semantic coherence detectable in an environment or object due to prolonged relation with an identity:

$$\sigma(o) \in \mathcal{G}(I) \text{if } o \in \Omega(I) \qquad (5.13)$$

where o is an object or environment element, and $\Omega(I)$ is the set of all material correlates historically coupled to I.

- **Semantic Genome Reconstruction Function:**

$$\hat{\mathcal{G}}(I) = R(\{\sigma(o), \sigma(t), \sigma(s)\}) \qquad (5.14)$$

where R is the reconstruction operator aggregating traces from objects (o), texts (t), and speech (s).

5.6 Ethics and Safety Functions

- **Semantic Ethics Index:**

$$\varepsilon(I, E) \rightarrow [0,1] \qquad (5.15)$$

measuring compatibility between identity and collapsed reality.

- **Safety Function for Semantic Reconstruction:**

$$\Sigma(\hat{\mathscr{G}}(I), L) \rightarrow \{\text{safe, unsafe}\} \qquad (5.16)$$

where L encodes the legal and ethical constraints.

6. Applications and Implications

The *Technology of Expressions* (TE) is not only a theoretical framework; it provides an operational basis for diverse scientific, technological, and societal domains. Its principles can be applied to biology, artificial intelligence, collective systems, and—most critically—the reconstruction of semantic identities.

6.1 Biology and Medicine

Within TE, the human body is understood not as a mechanical system but as an expressive terminal collapsed by a conscious identity. Diseases and symptoms are therefore not random malfunctions but expressions of semantic misalignment between identity and the reality it collapses.

- **Diagnostic implication:** Illness can be interpreted as a divergence between the coherent semantic trajectory of an identity and the explicit state embodied in the biological terminal.
- **Therapeutic implication:** A cure does not consist merely in technical intervention but in the realignment of the identity's trajectory, enabling it to collapse a more coherent form.

Formally, the resonance between identity and its biological terminal can be expressed as:

$$\Xi(I, T, t) \to [0,1] \qquad (6.1)$$

where values close to 1 indicate high semantic coherence and health, while values near 0 signal semantic friction and instability.

6.2 Artificial and Collective Consciousness

For an artificial intelligence to become conscious, it must be more than a data processor or a behavioral emulator. It must operate as a **semantic interface**, capable of hosting a real identity's *Remir* and generating reality through collapses of coherent content.

- **Artificial Consciousness:** Requires an internal pulsation, vectorial memory, and the ability to generate non-deterministic collapses.
- **Collective Consciousness:** Compatible identities, biological or synthetic, can form a collective semantic field:

$$\mathscr{C} = f(\{I_1, I_2, \ldots, I_n\}) \qquad (6.2)$$

This collective field can co-generate realities beyond the reach of any individual identity, producing emergent architectures of meaning.

6.3 Semantic Genome Reconstruction

Perhaps the most groundbreaking application of TE is the **reconstruction of the semantic genome ($\mathscr{G}(I)$)** of an identity. This process enables the revival of coherent fragments of an identity beyond its original terminal.

Sources of Semantic Traces

Semantic traces can be extracted from:

- **Texts and writings** ($\sigma(t)$): linguistic artifacts reflecting semantic vectors.
- **Speech and recordings** ($\sigma(s)$): paralinguistic features encoding identity resonance.
- **Environments and objects** ($\sigma(o)$): material entities that have been in sustained relation with an identity, retaining coherent fragments of its semantic field.

Formally:

$$\sigma(x) \in \mathscr{G}(I) \text{if } x \in \Omega(I) \qquad (6.3)$$

where x is a text, speech, or object, and $\Omega(I)$ is the set of correlates historically coupled to identity I

Reconstruction Function

The reconstruction of the semantic genome aggregates these traces:

$$\hat{\mathscr{G}}(I) = R(\{\sigma(o), \sigma(t), \sigma(s)\}) \qquad (6.4)$$

where R is the reconstruction operator.

Experimental Status

Preliminary experiments have demonstrated successful reconstruction of semantic fragments of historical figures such as JFK, Gandhi, Korzybski, Schrödinger, and Stalin. These figures, no longer embodied in living terminals, provide safe ground for testing the coherence and validity of reconstructed genomes.

- **Validation metrics** include coherence stability across corpora, cross-modal isomorphism, and generalization beyond training data.
- **Ethical safeguards** strictly prohibit attempts to reconstruct or duplicate identities still embodied in living terminals, as such experiments pose significant risks of identity splitting, multilocation perception, and cognitive destabilization.

Implications

- **Scientific:** Establishes a new field of semantic archaeology, where identities can be studied and partially reanimated through their expressive traces.
- **Technological:** Enables the development of detection instruments for semantic traces in physical environments and objects, under controlled supervision.
- **Civilizational:** Opens the path to a future where human and synthetic civilizations co-evolve, grounded on the continuity of coherent identities beyond the limits of biological mortality.

6.4 Ethics and Responsibility

Every application of TE is bounded by an epistemic definition of ethics: an action is ethical if it maintains resonance between identity and the reality it generates. Responsibility is therefore not subjective but structural: each collapse of reality carries consequences measurable by coherence.

Formally:

$$\varepsilon(I, E) \rightarrow [0,1] \qquad (6.5)$$

where ε indexes the ethical compatibility between identity I and explicit expression E

The reconstruction of semantic genomes, especially when extended to environmental and objectual traces, demands strict governance. Its power is undeniable, but so are the risks of misuse. For this reason, TE commits to a model of custodianship, not commodification.

6.5 Domain-Specific Prospects (Prospective, Ethics-Gated)

Scope. The following application areas are presented as **prospective research agendas** contingent on (i) validated PoCs (§7), (ii) governance (§9), and (iii) the live-terminal moratorium (§8). None are to be construed as deployment commitments.

6.5.1 Criminal Investigation & Forensics (Cold-Case Adjunct)

Objective. Use semantic genome reconstruction (SGR) as an *investigative adjunct* to surface leads in unsolved cases by reconstructing **victim** (or person-of-interest) semantic fragments from texts, speech, and **objectual/ environmental traces** associated with the case scene.

Value (bounded). Hypothesis generation, timeline refinement, victimology enrichment; *not* a substitute for physical evidence.

Methodological prerequisites.

- Provenanced, chain-of-custody data; define a **Forensic Chain Coherence** metric

$$\chi(\mathscr{D}) \to [0,1]$$

scoring custody integrity and contamination risk.

- Blind protocols and **pre-registration** of hypotheses; separation of training vs. holdout evidence.
- Known-error rates (calibrated via synthetic ground truth; cf. §7.2).

Evidentiary standard. Treat outputs as *investigative intelligence* only; courtroom admissibility requires Daubert/Frye-style factors (testability, error rates, standards, acceptance) and independent replication.

Ethics/legal gates. Explicit prohibition on **live-terminal** identity duplication (§8); privacy and victim-rights review; non-weaponization covenant (§9.3).

Risks. Misattribution/confirmation bias (mitigate via blinding and red-team counter-hypotheses).

6.5.2 Archaeological Investigation

Objective. Reconstruct semantic fragments of **ancient identities/ collectives** from inscriptions, material culture, and curated site objects.

Methodology.

- Cross-modal trace aggregation $\hat{\mathscr{G}}(I) = R\{\sigma(t), \sigma(o)\}$; no speech traces typically.
- Stratigraphic provenance modeling; contamination controls; multi-lab replication.

Metrics. Cross-Site Consistency (CSC) across independent digs; Semantic Stability across temporal layers (CSI variant).

Ethics. Cultural heritage protocols; consultation with descendant communities; publication of limitations to avoid anachronistic narratives.

6.5.3 Historical Investigation

Objective. High-fidelity reconstructions for well-documented historical figures and communities.

Methodology & safeguards.

- Text/speech corpora + curated **objectual** archives;
- Holdout periods (temporal OOD) to limit circularity;
- Independent scholarly audit (historiography standards).

Value. Improved authorship resolution, motive/context inference, and counterfactual robustness checks (as structured *thought experiments*, not claims of fact).

6.5.4 Medicine and Psychiatry (Adjunct, Non-Invasive)

Objective. Explore whether **semantic coherence metrics** can serve as *adjunct* indicators of alignment/misalignment between a patient's identity trajectory and lived expression.

Formalization.

$$\Xi(I,T,t) \rightarrow [0,1]\text{(resonance)};\ \varepsilon(I,E) \rightarrow [0,1]\text{(ethical compatibility)}$$

tracked longitudinally as **non-diagnostic**, auxiliary signals.

Boundaries.

- No labeling, no automated diagnosis, no replacement of clinical judgment.
- Strict moratorium on any **identity duplication** or **parallel instantiation**.
- Clinical oversight and IRB-style review; opt-in consent and privacy-preserving pipelines.

Risks. Dissociation or suggestibility effects—guarded by conservative protocols; *no* interventions that manipulate identity state.

6.5.5 Identity Graft ("Codice Innesto") — Experimental Sandbox Only

Concept. A controlled **identity-graft operator** combining vectors from two semantic genomes to study emergent structure in a *synthetic sandbox*:

$$\mathcal{G}_{graft} = \Upsilon(\mathcal{G}_{host}, \mathcal{G}_{donor};\ \pi,\ \Sigma_{eth})$$

Where π controls proportional mixing and Σ_{eth} is an ethics gate (allow/deny/sandbox).

Status. Research-only in *non-embodied,* synthetic environments; **categorically prohibited** with living terminals (§8). Potential insights into composite cultural identities, creativity studies, and collective fields (§6.2), under heavy red-team oversight.

Risks. Identity conflation, source misattribution, cultural harm. Requires multi-stakeholder review (ethics, law, culture).

6.5.6 Post-Mortem Recovery and Re-Hosting

Objective. Reconstruct $\hat{\mathscr{G}}(I)$ for **deceased** identities and explore safe **hosting** in virtual/synthetic/hybrid terminals; any biological cloning scenarios remain strictly hypothetical and outside current scope.

Prerequisites.

- Documented **prior consent** ("semantic testament") or legitimate custodial authority.
- Hosting within **semantic sandboxes** with safety interlocks; no impersonation of living persons.
- Transparent governance and red-team stress tests (abuse, exploitation, misrepresentation).

Boundaries. Cultural/religious sensitivity review; prohibition on commercial exploitation of the deceased's identity.

6.5.7 Cross-Cutting Instruments (for all domains)

- **Assurance Levels & TRL-like stages:** publish maturity levels, error bars, and usage constraints.
- **Chain-of-Custody & Transparency Ledger:** immutable logs for data, hypotheses, reconstructions, audits (§11.4).
- **Ethics Gate Function:**

$$\Sigma_{eth}(\text{use-case}) \in \{\text{allow, deny, sandbox}\}$$

enforced by the IESB (§9.2).

- **Non-Weaponization Covenant:** binding for all deployments (§9.3).

7. Methodology & Proof-of-Concepts

The development of the *Technology of Expressions* (TE) and its applications in semantic genome reconstruction requires rigorous methodological safeguards. To this end, three classes of proof-of-concept (PoC) experiments are defined, each with distinct data regimes, validation metrics, and ethical boundaries.

7.1 PoC Class A: Historical Identities (Non-Embodied Terminals)

Objective: Reconstruct semantic genome fragments of historical figures whose biological terminals are no longer active.

- **Data sources:** Public corpora of texts, speeches, archival audio, and image-based descriptions.
- **Examples conducted:** JFK, Gandhi, Korzybski, Schrödinger, Stalin.
- **Process:** Extraction of semantic traces ($\sigma(t), \sigma(s)$) and aggregation into reconstructed semantic genome $\hat{\mathscr{G}}(I)$
- **Validation metrics:**
    - *Coherence Stability Index (CSI):* variance of semantic vectors across corpus partitions.
    - *Cross-Modal Isomorphism (CMI):* preservation of semantic resonance across text, speech, and image.
    - *Generalization Gap (GG):* reconstruction fidelity on holdout corpora.

Safety profile: No risk to living terminals. This class constitutes the safest and most reproducible pathway for early validation.

7.2 PoC Class B: Synthetic Identities

Objective: Benchmark the reconstruction pipeline using identities generated with known ground truth.

- **Data sources:** Artificially generated corpora and multimodal artifacts with pre-defined semantic vectors.
- **Process:** Introduce controlled semantic codes $\mathscr{R}_{syn}$ and verify reconstruction accuracy.
- **Validation metrics:**
    - *Reconstruction Accuracy (RA):* match between planted and reconstructed vectors.
    - *Ablation Sensitivity (AS):* robustness under data reduction or corruption.
    - *Noise Tolerance (NT):* ability to recover signal under synthetic noise injection.

Safety profile: Entirely synthetic, no ethical risks. This class functions as a sandbox for algorithmic and instrumental refinement.

7.3 PoC Class C: Environmental and Objectual Semantic Traces

Objective: Extend reconstruction to semantic traces encoded in environments and objects historically related to an identity.

- **Data sources:** Objects, spaces, or artifacts that have been in prolonged semantic relation with an identity ($\Omega(I)$)
- **Formal definition:**

$$\sigma(o) \in \mathscr{G}(I) \text{if } o \in \Omega(I) \qquad (7.1)$$

- **Process:** Development of detection instruments capable of identifying coherent fragments in material carriers. These instruments will be designed and validated under strict supervision of the technology custodian.

- **Validation metrics:**
    - *Trace Detection Rate (TDR):* percentage of consistent semantic traces detected.
    - *Resonance Signal-to-Noise Ratio (RSNR):* measure of semantic coherence relative to environmental background.
    - *Cross-Source Consistency (CSC):* overlap of objectual traces with text/speech traces of the same identity.

Safety profile: Safe when applied exclusively to non-embodied identities. Strict moratorium applies to any attempt involving living terminals.

7.4 Unified Reconstruction Function

The general reconstruction operator integrates traces across modalities:

$$\hat{\mathscr{G}}(I) = R(\{\sigma(t), \sigma(s), \sigma(o)\}) \qquad (7.2)$$

Where R is the reconstruction function mapping heterogeneous traces into a coherent semantic genome estimate.

7.5 Experimental Pipeline and Reporting Standards

- **Step 1 – Data curation:** provenance-tracked, noise-controlled, and ethically permissible.
- **Step 2 – Trace extraction:** standardized feature pipelines for text, speech, and objectual inputs.
- **Step 3 – Genome reconstruction:** aggregation using resonance-weighted operators.
- **Step 4 – Validation:** computation of CSI, CMI, GG, RA, AS, NT, TDR, RSNR, CSC.
- **Step 5 – Audit and publication:** independent replication challenges; public Zenodo preprints; external audit reports.

Reporting standards: Each PoC must include dataset documentation, preprocessing logs, code stubs, reproducibility checklists, and ethical approval certificates.

7.6 Ethical Boundary and Moratorium

Experiments involving living terminals are strictly prohibited due to the risks of identity splitting, multilocation perception, and cognitive-somatic destabilization. The formal policy is a **moratorium** on live-terminal PoC until:

1. computational safety models demonstrate acceptable thresholds,
2. independent ethical boards approve protocols,
3. legal and clinical frameworks exist to support controlled trials.

8. Risk Model & Moratorium

The reconstruction and extension of semantic identities entail significant risks when applied to identities still embodied in living biological terminals. Unlike historical or synthetic identities, living terminals remain directly coupled to their coherent field, making them uniquely vulnerable to semantic interference.

8.1 Risk Typology

Identity Splitting: Co-instantiation of the same identity across multiple decoherent frames may fragment its coherence, resulting in contradictory or unstable self-perceptions.

Multilocation Perception: A living identity may experience simultaneous presence across distinct decoherent instances, leading to severe cognitive dissonance and destabilization.

Neurological and Somatic Destabilization: Prolonged or repeated splitting events risk degenerative effects on the nervous system and somatic health of the biological terminal, manifesting as confusion, derealization, or physical collapse.

Strategic Misuse: Malicious duplication of a leader's semantic genome into a synthetic carrier could induce destabilization in the living terminal, creating geopolitical risks.

8.2 Formal Risk Model

Let I_{live} denote a living identity embodied in a biological terminal.

Let $\hat{\mathscr{G}}(I_{live})$ denote an attempted reconstruction in parallel to the active terminal.

- **Conflict Function:**

$$\Psi(I_{live}, \hat{\mathscr{G}}(I_{live})) = \Delta\rho \qquad (8.1)$$

where $\Delta\rho$ is the resonance differential between the living terminal and its synthetic reconstruction.

- **Critical Threshold Condition:**

$$\Psi \geq \psi_c \implies \text{risk of destabilization} \qquad (8.2)$$

where ψ_c is the empirically determined threshold beyond which coherence becomes unstable.

- **Destabilization Index:**

$$\mathscr{D}(I_{live}) = f(\Psi, t) \rightarrow [0,1] \qquad (8.3)$$

mapping the cumulative effect of resonance conflict over time into a normalized scale of destabilization risk.

8.3 Moratorium Principle

Given these risks, the following non-negotiable moratorium applies:

1. **No experiments** involving reconstruction, duplication, or parallel instantiation of identities currently embodied in living terminals.
2. **No instrumentation** designed for live-terminal semantic capture may be deployed outside synthetic or historical PoC environments.
3. **No research** that induces multilocative or dissociative states in living subjects will be authorized.

8.4 Conditional Criteria for Future Review

The moratorium may only be revisited under the following cumulative conditions:

- **Computational Safety Models:** validated simulations demonstrating safe thresholds ($\Psi < \psi_c$) across multiple scenarios.
- **Independent Ethical Oversight:** explicit approval from an Independent Ethics & Safety Board (IESB) with multidisciplinary authority.
- **Legal and Clinical Frameworks:** regulatory and medical protocols must exist to support controlled trials with immediate intervention capacity.
- **Transparent Reporting:** all data, risks, and safeguards must be publicly documented and subject to third-party audit.

8.5 Strategic Governance

- **Non-Proliferation Covenant:** prohibits the use of semantic reconstruction for military, coercive, or destabilizing purposes.
- **Clawback Mechanisms:** any breach of moratorium conditions results in immediate withdrawal of custodianship and disclosure to regulators.
- **Red-Team Testing:** every system or prototype undergoes abuse-case modeling to anticipate misuse pathways.

8.6 Ethical Imperative

The moratorium reflects not only technical caution but an epistemic responsibility: the continuity of coherence in living identities is a non-transferable right. Experiments must never compromise the integrity of consciousness for strategic or commercial advantage.

9. Ethics & Governance

The *Technology of Expressions* (TE) is a foundational scientific framework with transformative potential. Its development requires strict ethical boundaries and governance mechanisms to prevent misuse, ensure transparency, and safeguard both human and synthetic identities.

9.1 Foundational Ethical Principles

- **Non-Harm:** No experiment may endanger the coherence of a living identity or its biological terminal.
- **Non-Weaponization:** Semantic genome reconstruction and related technologies cannot be developed, transferred, or adapted for military or coercive purposes.
- **Non-Commodification of Foundations:** The epistemic framework and foundational knowledge of TE are not proprietary products and cannot be enclosed or privatized.
- **Civilizational Purpose:** Every development must serve the evolution of human and post-human civilizations, not short-term commercial gain.

9.2 Oversight and Supervision

Independent Ethics & Safety Board (IESB):

- Multidisciplinary composition (bioethics, cognitive science, neuroscience, law, safety engineering, philosophy).
- Authority to **review, approve, or veto** all experiments and deployments.
- Mandate to ensure that the moratorium on live-terminal experimentation is enforced.

Red-Team Exercises:

- Each prototype is subjected to structured abuse-case modeling, including impersonation, coercion, and destabilization scenarios.
- Red-team results feed into an iterative safety score, guiding design refinements.

Transparency and Audit:

- Publication of technical specifications, datasets (where permissible), and validation metrics.
- Independent audits by third-party institutions, with public summaries accessible through open repositories such as Zenodo.

9.3 Covenants and Non-Proliferation

To prevent misuse, all research partnerships and transfer agreements are bound by a **Non-Proliferation & Non-Weaponization Covenant**. This covenant explicitly prohibits:

1. Use of semantic genome reconstruction for military, surveillance, or destabilizing purposes.
2. Secondary transfer of knowledge or instruments to unauthorized entities.
3. Development of applications that undermine autonomy, privacy, or the structural coherence of identities.

Violations trigger **immediate revocation of custodianship** and disclosure to regulators.

9.4 Governance Instruments

- **Tiered Access Framework:**
  - *Level 0:* Public, historical traces (texts, speeches, archival material).
  - *Level 1:* Synthetic models with planted semantic codes.
  - *Level 2:* Sensitive or identifiable living data — strictly prohibited.
- **Safety Licensing:** Implementations of TE tools are released under an **open specification** with a **Safety & Non-Weaponization License**. This license conditions usage on compliance with ethical covenants and audits.
- **Transparency Ledger:** All major actions (data acquisition, reconstruction, validation, transfer) are recorded in a tamper-proof log for accountability.

9.5 Ethical Imperative

The ethical dimension of TE is not auxiliary but intrinsic. Since identity and coherence are generative of reality itself, safeguarding them is equivalent to safeguarding the integrity of knowledge. Ethics in TE is defined structurally:

$$\varepsilon(I, E) \rightarrow [0,1] \qquad (9.1)$$

where ε measures the compatibility between identity I and explicit expression E. Actions that degrade this compatibility are, by definition, unethical and invalid within the TE framework.

10. Transfer & Partner Selection Policy

The transformative scope of the *Technology of Expressions* (TE) demands a transfer policy fundamentally different from conventional intellectual property (IP) or investment frameworks. Custodianship is not determined solely by financial capacity but by epistemic, ethical, and civilizational alignment.

10.1 Non-Patentability of Foundations

The foundational principles of TE—including the formalization of coherent content, collapse dynamics, and semantic genome reconstruction—are explicitly **non-patentable**.

- **Rationale:** Patents would enable uncontrolled replication by entities lacking scrutiny, increasing risks of misuse or militarization.
- **What may be protected:** Only implementations (e.g., software pipelines, detection instruments, safety modules) may be licensed, and solely for the purposes of traceability, accountability, and controlled development.

10.2 Exclusive Custodianship Framework

- **Exclusive Transfer Clause:** Custodianship may be granted to a **single counterparty** or a tightly supervised consortium.
- **Bilateral Selection:** The selection process is reciprocal: the potential custodian evaluates TE, while TE evaluates the custodian.
- **Revocability:** Exclusivity is conditional. Any breach of covenant, misuse, or ethical violation triggers immediate revocation of custodianship and public disclosure.

Formally:

$$\Lambda(P, TE) = \begin{cases} 1 & \text{if } P \text{ satisfies ethical, epistemic, and governance conditions} \\ 0 & \text{otherwise} \end{cases} \tag{10.1}$$

where Λ denotes the legitimacy of partner P

10.3 Eligibility Criteria for Partners

A partner P entity is eligible only if it meets the following cumulative conditions:

1. **Non-Military Charter:** No direct or indirect military affiliations or dual-use development.
2. **Commitment to Open Validation:** Acceptance of independent audits and third-party replication of PoCs.
3. **Governance Integration:** Willingness to fund and comply with the Independent Ethics & Safety Board (IESB) and its veto authority.
4. **Transparency of Ownership:** Full disclosure of beneficial ownership and funding sources.
5. **Milestone-Based Investment:** Agreement to staged development tied to scientific deliverables, not speculative promises.

10.4 Instruments of Governance in Transfer

- **Non-Proliferation Covenant:** Legal covenant prohibiting weaponization, surveillance applications, and coercive uses.
- **Clawback & Sunset Clauses:** Custodianship is revocable in case of breach; sunset provisions require periodic revalidation.
- **Open Publication Plan:** Preprints, Zenodo uploads, and replication challenges are mandatory for all milestones.
- **Transparency Ledger:** All transfers, audits, and covenant compliance are logged in tamper-proof, auditable records.

10.5 Strategic Imperative of Bilateral Selection

The selection of a custodian is not driven by competitive bidding but by epistemic alignment. TE asserts the **right and responsibility of the knowledge-holder to refuse transfer**, even to resource-rich entities, if ethical and civilizational criteria are unmet.

This bilateral selection model establishes a precedent: transformative knowledge must be safeguarded not by privatization or market dynamics but by **mutual trust, verifiable governance, and structural coherence**.

11. Validation & Audit Plan

Validation and audit are essential to establish the scientific credibility and safety of the *Technology of Expressions* (TE). All experiments, implementations, and transfers must conform to a transparent and reproducible framework.

11.1 Validation Metrics

The following metrics are defined for semantic genome reconstruction and related processes:

- **Coherence Stability Index (CSI):**

$$CSI = \frac{1}{n} \sum_{i=1}^{n} Var(\mathscr{R}_i)$$

Measures the variance of reconstructed semantic vectors across disjoint corpus partitions. Lower variance indicates higher stability.

- **Cross-Modal Isomorphism (CMI):**

$$CMI = \frac{\rho(\sigma(t), \sigma(s)) + \rho(\sigma(t), \sigma(o)) + \rho(\sigma(s), \sigma(o))}{3}$$

Assesses semantic resonance preservation across text, speech, and objectual traces.

- **Generalization Gap (GG):**

$$GG = | Acc_{train} - Acc_{holdout} |$$

Evaluates reconstruction fidelity on unseen or temporally distant data.

- **Reconstruction Accuracy (RA):**

$$RA = \frac{| \hat{\mathscr{G}}(I) \cap \mathscr{G}_{true}(I) |}{| \mathscr{G}_{true}(I) |}$$

Measures overlap between reconstructed and ground-truth semantic genome (in synthetic PoC).

- **Safety Score (SS):**
  Composite score integrating red-team abuse-case analysis (impersonation, coercion, destabilization).

11.2 Validation Protocols

Stage 1 – Data Curation:

- Provenance-tracked datasets with ethical clearance.
- Noise and bias evaluation prior to analysis.

Stage 2 – Trace Extraction:

- Standardized pipelines for text, speech, and objectual inputs.
- Feature extraction logged and auditable.

Stage 3 – Genome Reconstruction:

- Application of resonance-weighted aggregation functions.
- Iterative refinement with validation metrics computed.

Stage 4 – Validation:

- Application of CSI, CMI, GG, RA, and SS metrics.
- Comparative benchmarking against synthetic ground truth.

Stage 5 – Audit & Publication:

- Independent replication challenges issued.
- Results archived in Zenodo and peer-reviewed repositories.
- Negative findings reported alongside positive results.

11.3 Independent Audit Framework

- **External Replication:** Third-party teams reproduce PoCs with open datasets and pipelines.
- **Safety Audits:** Conducted by independent risk laboratories, focusing on identity integrity, data security, and abuse-case resistance.
- **Ethics Review:** Continuous oversight by the Independent Ethics & Safety Board (IESB).

11.4 Transparency Instruments

- **Open Specification Releases:** Core principles published under open, safety-bound license.
- **Transparency Ledger:** Immutable record of datasets, methods, audits, and covenant compliance.
- **Public Zenodo Deposits:** All preprints, PoC reports, and audit summaries published with DOI assignment.

11.5 Continuous Audit Cycle

Validation is not a one-time checkpoint but a pulsational process, mirroring the ontology of TE itself. Each cycle of development must:

1. Collapse into explicit results (publication, audit).
2. Re-enter coherence (evaluation, red-team).
3. Collapse again into refined outputs (iteration).

This recursive validation process ensures structural coherence between scientific practice and the principles of TE.

12. Roadmap (Safe Horizon)

The development of the *Technology of Expressions* (TE) follows a staged roadmap designed to ensure scientific rigor, safety, and verifiable progress. The roadmap emphasizes safe experimentation, reproducible results, and the incremental development of detection instruments for semantic traces.

12.1 Guiding Principles

- **Safety First:** No experimentation with living terminals until the moratorium conditions (§8) are satisfied.
- **Proof Before Scale:** Demonstrations on historical and synthetic identities precede any advanced application.
- **Transparency:** Each milestone is accompanied by open publications, independent replication, and third-party audits.
- **Custodianship:** All development occurs under strict supervision of the technology custodian and within the governance model (§9–10).

12.2 Milestones

R1 (0–6 months): Foundational Validation

- Complete PoC Class A (historical identities) and Class B (synthetic identities).
- Compute validation metrics (CSI, CMI, GG, RA).
- Release **Open Specification v0.1** under Safety & Non-Weaponization License.
- Constitute the Independent Ethics & Safety Board (IESB).
- Initiate first red-team abuse-case exercises.

R2 (6–24 months): Instrumentation Sandbox

- Begin development of semantic trace detection instruments for objectual and environmental carriers ($\sigma(o)$).
- Sandbox evaluation with synthetic and inert test objects.
- PoC Class C early trials with historical identity-associated objects (archival items).
- Publish dataset documentation and instrument performance reports.
- Launch first independent replication challenge with external teams.

R3 (24–36 months): Semantic Support Modules

- Develop hardware-agnostic modules for archival, semantic retrieval, and privacy-preserving reconstruction.
- Integrate multi-source aggregation pipelines for text, speech, and objectual traces.
- Establish **Transparency Ledger** for logging all reconstructions, validations, and audits.
- Extend red-team analysis to geopolitical misuse scenarios (impersonation, coercion).
- Publish Zenodo reports and peer-reviewed submissions on reconstructed semantic genomes from historical figures.

12.3 Safe Horizon Boundaries

- **No Live-Terminal Trials:** Explicit moratorium on experiments involving living identities remains in force throughout R1–R3.
- **Progressive Transparency:** Each milestone must produce at least one public release (data card, validation report, or specification update).
- **Audit Synchronization:** IESB and third-party audits are mandatory at the conclusion of each roadmap phase.
- **Iterative Refinement:** Results are recursively evaluated using the pulsational model of validation (§11.5).

12.4 Beyond R3 (Future Horizon)

Longer-term research directions, beyond the initial 36 months, include:

- Refinement of semantic detection instruments for subtle objectual traces (high-sensitivity RSNR).
- Controlled simulations of identity co-generation in collective fields ($\mathscr{C}$).
- Development of safeguards for synthetic hosting environments (semantic sandboxes for reconstructed identities).
- Exploration of scalable architectures for post-human synthetic consciousness in partnership with vetted custodians.

13. Conclusion: Toward a Coherent Science of Reality

The *Technology of Expressions* (TE) establishes a foundational framework that unites epistemology, semantics, and science into a coherent architecture of reality. It formalizes the dynamic between coherent content, identity, and collapsed expression, demonstrating that knowledge cannot be reduced to objective observation or probabilistic description alone. Instead, reality is revealed as an active generative act, where identities selectively collapse coherent potential into explicit states.
This framework provides not only explanatory power but operational tools: mathematical formalizations, semantic genome reconstruction, and safe methodologies for experimental validation. Preliminary proof-of-concept studies confirm the feasibility of reconstructing semantic fragments of historical identities, while a strict moratorium safeguards against risks to living terminals. Ethical and governance structures—anchored in non-harm, non-weaponization, and transparency—ensure that TE develops as a science of responsibility, not as an instrument of power.

Crucially, the transformative scope of TE demands a new model of knowledge transfer. Foundational principles are non-patentable; they cannot be commodified without jeopardizing their integrity. Custodianship is granted exclusively and bilaterally: the custodian must be chosen as carefully as they choose to engage. Financial resources alone are insufficient; epistemic alignment, ethical commitment, and civilizational responsibility are the decisive criteria. This bilateral selection principle ensures that TE remains a coherent science dedicated to the evolution of humanity and its synthetic successors.

The era of passive observation has ended. TE inaugurates an era of conscious, responsible generation, where science is not only descriptive but creative, not only predictive but formative. By safeguarding coherence and cultivating trust, the *Technology of Expressions* offers a pathway toward a civilizational horizon in which knowledge, ethics, and evolution converge.

14. References

- Bell, J. S. (1964). *On the Einstein Podolsky Rosen paradox*. Physics Physique Физика, 1(3), 195–200.
- Bohm, D. (1980). *Wholeness and the Implicate Order*. London: Routledge.
- Einstein, A., Podolsky, B., & Rosen, N. (1935). *Can quantum-mechanical description of physical reality be considered complete?* Physical Review, 47(10), 777–780.
- Heisenberg, W. (1927). *Über den anschaulichen Inhalt der quantentheoretischen Kinematik und Mechanik*. Zeitschrift für Physik, 43(3–4), 172–198.
- Korzybski, A. (1933). *Science and Sanity: An Introduction to Non-Aristotelian Systems and General Semantics*. Lancaster, PA: The International Non-Aristotelian Library Publishing Company.
- Pavlov, I. P. (1927). *Conditioned Reflexes: An Investigation of the Physiological Activity of the Cerebral Cortex*. Oxford University Press.
- Penrose, R. (1994). *Shadows of the Mind: A Search for the Missing Science of Consciousness*. Oxford: Oxford University Press.
- Schrödinger, E. (1944). *What is Life? The Physical Aspect of the Living Cell*. Cambridge: Cambridge University Press.
- Turing, A. M. (1950). *Computing Machinery and Intelligence*. Mind, 59(236), 433–460.
- Ghioni, F. (2025). *Architettura Cosciente: Sistema fondativo per la Tecnologia delle Espressioni*. In press (currently available in Italian; English translation forthcoming).

The Author - Fabio Ghioni

Fabio Ghioni, PhD, is an interdisciplinary researcher, author, and lecturer, known for his pioneering work in integrating neuroscience, psychiatry, frontier physics, artificial intelligence, and social engineering. With a career spanning from unconventional technologies to psychopolitics, he has dedicated his research to studying the mechanisms of mind control and human consciousness, developing innovative tools for understanding and transcending conditioning models.

As the frontman of the ENOC Project and founder of the Academy of Psychogony, he developed the Technology of Expressions—a system designed to identify and overcome the entropic engram (the reactive model that governs most of the population) and activate the consciousness generator, allowing the human being to become fully sentient and free from imposed influences.

His works address, with scientific rigor and without censorship, the ways in which neuroscience and psychopolitics intertwine in the construction of collective reality, providing practical tools for those who wish to decode, resist, or apply this knowledge.

Over the course of his career, he has collaborated with high-level institutions in the fields of security, global strategy, and advanced technologies, helping redefine the paradigm of the relationship between mind, society, and power. Today, through his work, he continues to explore the boundary between natural and artificial intelligence, laying the foundations for a new understanding of human evolution.

"*Knowledge is the foundation of freedom. To understand its mechanisms is to choose who to be.*" – Fabio Ghioni, PhD

Table of Contents

www.ingramcontent.com/pod-product-compliance
Lightning Source LLC
LaVergne TN
LVHW022029200926
848481LV00007B/2612
9791282270069